Casting Premium Bullets for the Black Powder Cartridge Rifle

Casting Premium Bullets for the Black Powder Cartridge Rifle

by

Paul A. Matthews

Wolfe Publishing Company

2180 Gulfstream, Ste. A
Prescott, Arizona 86301

ISBN: 978-1-879356-58-0

22 21 20 19 18 5 4 3 2 1

Published May 2001

Reprinted February 2017

Printed in the United States of America

Cover Design: Gerald Hudson
Production Design: Becky Pinkley

Dedication

To my father and mother who have been gone for many years but who left me a heritage rich in free thinking and expression, a heritage encompassing many patriots who fought in the American Revolution and reaching on back to eight brave souls who challenged the Atlantic on the decks of the *Mayflower*. To my father and mother who gave me all this and much more.

To the Shooter

The home-grown cast bullet carries with it from the muzzle to the target the sum total of our skills as casters of bullets, loaders of ammunition and our proficiencies with rifles.

– Matthews

Contents

WARNING

Casting bullets is a dangerous operation that can easily result in severe burns, disfigurement and/or lead poisoning. Despite the fact that many of us have cast bullets accident free for half a century or better, and despite the fact that in years past millions of tons of lead have been cast into bullets by the individual as well as by military units and manufacturers, the fact remains that a single moment of carelessness during the casting process can get you or someone in close proximity severely burned, disfigured or even killed.

Since neither I nor the publisher of this book have any control over the quality of your casting equipment, or the procedures you use before, during and after casting, we hereby disclaim any responsibility for any damage, injury or death that may be incurred as a result of using any part of or all of the information contained in this book.

If you choose to cast bullets, you are strictly on your own and are solely responsible for any and all accidents, mishaps, injuries and damage to health or property incurred as a result of casting bullets.

Preface

I cast my first bullet back in 1949 using an old cast iron kettle on a kerosene stove and with a bent teaspoon for a dipper. Later, Charlie Canoll was being kind when he scanned those 400 plus bullets, shook his head and said, "If we look them over, we might find five or six worth shooting!"

Since that long ago day, I have cast tens of thousands of bullets from dozens of different moulds using everything from pure lead up through the lead-tin alloys to Linotype, monotype, printing plates and even zinc. After leaving the kitchen and the kerosene stove to my wife, I moved operations into an abandoned chicken coop and built a coal furnace out of a 50-pound pail, a 10-quart pail, some sheetmetal and lots of asbestos. With that, I could melt my bullet metal into something softer than slush.

The coal-fired furnace was soon replaced by a gasoline plumber's furnace from Sears, Roebuck & Company that served me for many years until I bought a Lyman electric furnace with a bottom pour spout. The vise-squeezed teaspoon had long since been discarded in favor of one of Lyman's time honored bullet casting ladles.

I used the Lyman electric furnace until the thermostat died, sent it home for a new thermostat and used it until *that* thermostat died. Then I did some rewiring to bypass the dead thermostat and continued to use that electric furnace until the heating element ruptured and gave up the ghost.

Somewhere along the line I learned that hand-dipped bullets were far superior to the bottom pour variety, so I removed the valve from the Lyman furnace and drove a steel ball bearing into the bottom pour nozzle. From that day to this, I have used the dipper.

During the latter days of the Lyman electric furnace, I also used two different Coleman stoves with Coleman fuel and the usual Lyman cast iron melting pot.

All this was fine except for one thing – metal capacity. At the time it did not bother me that after casting a few dozen bullets I had to replenish the pot. Usually after using a pot full of bullet metal, I was ready to quit anyhow. But when I got into the black powder silhouette shooting and needed bullets by the hundreds – yes, thousands – and I learned that replenishing the pot changed the temperature of my bullet metal along with the weight of my bullets – then I knew I had to have something bigger, something that would let me hold the temperature to plus or minus 10 degrees Fahrenheit while casting at least 200 .45-caliber bullets weighing up to 550 grains apiece.

My something bigger turned out to be a rugged 36,000 BTU propane burner sold by Gander Mountain and originally manufactured as Model 99PK by Metal Fusion, Inc. of Jefferson, Louisiana. This burner is made primarily as an outdoor cooker but works very well as a heat source for casting bullets. For a melting pot, I use a 2-quart cast iron sauce pan that holds about 35 pounds of bullet metal. I also have a 3-quart cast iron sauce pan that I have yet to use as the 2-quart pot seems fully adequate.

I should point out that in order to get and hold the temperature of the bullet metal at 820 to 850 degrees Fahrenheit, I use a sheetmetal wind screen around the melting pot. While there is no wind involved in my casting area, the wind screen holds the heat in around the melting pot. The wind screen is nothing more nor less than a strip of 3-inch wide sheetmetal bent into a circle 7¾ inches in diameter and the ends riveted together.

Metal Fusions, Inc. offers a number of other propane burners (outdoor cookers), among them being a 170,000 BTU double-burner unit. Jerry Matalavage uses one of these for his casting purposes and uses a heavy cast iron dutch oven as a melting pot!

This new propane burner took a little getting used to as far as heat regulation was concerned, but once mastered I found I could run anywhere from 200 to 250 superb bullets in about three and a half hours. I keep a thermometer in the bullet metal at all times, and the temperature seldom varies more than plus or minus 10 degrees during an entire casting session.

This is a long way from that first time in 1949 when I cast bullets in that old Ideal 3118 .32-20 mould. And where I was happy if I could hit a crow at 50 yards with the Ideal 3118 bullet, now it is my expectation to hit metallic turkey silhouettes on a regular basis at more than eight times that range.

Cast bullets can be superbly accurate, and if you plan to compete in the metallic silhouette matches, or in the long-range paper matches, your bullets *must* be superbly accurate. The first step in this direction is to cast a premium quality bullet – not just a so-so bullet but a premium quality bullet.

1

Do I Need Premium Cast Bullets?

The answer depends to a large extent upon how serious you are about shooting top scores. One gentleman I know, an excellent rifleman, pays little attention to the minor defects on the cast bullets he shoots. He claims that minor defects on the bases of cast bullets have little effect on their performance.

Many are the times I have sat behind the firing line watching this man shoot and wondered if this or that near miss on a turkey or ram was due to one of his not-so-perfect bullets? Many are the times I have watched him shoot and wondered if I or the rest of the shooters on the line could compete with him if he put forth that little bit of extra effort to make certain that every bullet he fired was of premium quality.

Believe me, when I first got into this black powder cartridge rifle silhouette game, I hated to throw bullets back into the lead pot because of some minor defect that probably would not put the bullet more than a fraction of an inch off its course. But as the years passed and my scores kept improving, I came to realize that perfection of the bullet was all-important, that a fraction of an inch could make the difference between a hit or a miss.

Think about it. If you are using a .45-70 throwing a 500- to 550-grain bullet at 1,200 fps, if you get *half the diameter* – .229 inch – on the ram, you are going to knock that ram from its stand! And that could be the point to win the match.

While you are the one who has to decide whether premium bullets are worth the extra effort, I can promise you that when I load for a match today, I do not load any bullet that is not in the premium category. Before any of my bullets gets lubricated, it is visually examined for even the most minor defect. Again after lubrication the base of each bullet is reexamined, and only bullets with perfect bases get seated in the cartridge case.

Does it pay off? You bet! With premium bullets and a well-developed powder charge, if you do your part behind the butt, you soon reach the point where if the bullet is off the target due to your hold, you know where it is off the target before the spotter does. And if the bullet is off the target or just out of the area of normal impact due to light change, you *know* you can make a sight adjustment and the next bullet will be where you want it! Premium bullets pay off!

2

What Is a Premium Cast Bullet?

Simply put, a premium cast bullet is one that has no defects, visible or invisible. It is a cast bullet with a sharp edge all around its base, a bullet whose base is absolutely smooth with no protrusion or cavitation at the cutoff. It is a cast lead alloy bullet whose driving bands are sharp and clearly defined with no rounded edges or minute imperfections caused by air being trapped in the mould. It is a bullet that when cast is as near like its predecessor and its successor as is humanly possible to obtain when using the same methods and much the same equipment as was used more than a century ago. It is a bullet whose only variation from its predecessor or successor is one of weight, very slight at that and caused by a slight change in the temperature of the melt or of the mould. The premium cast bullet is what you want to be shooting when you take your place on the firing line and look through the sights at those distant targets!

Why cast anything less than a premium bullet? Why not maintain a constant temperature of the melt and mould and thus decrease the amount of weight variation from bullet to bullet? It does not cost any more money or take much more time to cast a premium bullet as compared to a less than premium bullet.

Why not lay each bullet in the order cast upon a casting board (described in another chapter) instead of shoving them together in a pile where the hot, still soft bullet metal at the edge of the base can be deformed? Why not lay each bullet in the order cast upon the casting board to later be segregated by weight if that is your procedure, or to be placed in pistol cartridge boxes in the order cast to be loaded and fired still in the order cast?

If you have done your part during the casting process, you will find your bullets will not vary in weight much more than 1.5 grains from the lightest to the heaviest even when casting bullets as heavy as 550 grains. If your first few bullets weigh lighter than the norm, it is usually because the mould was not up to the normal casting temperature. Conversely, if the first few bullets weigh heavier than the norm, it is usually because the mould was too hot when you first started casting.

One thing other than temperature that will cause a change in bullet weight during the casting process is not fully seating the mould blocks together each time before casting. This is vital. If the mould blocks are not fully seated, the bullet being cast will be of slightly larger diameter and one or two grains heavier than a bullet cast when the blocks are fully seated.

In order to assure the mould blocks are fully closed and seated before casting a bullet, always, after closing the mould, tap the jaw of the handle adjacent to one of the mould blocks two or three times in the direction that tends to seat the blocks together. Always make certain the mould blocks are closed before tapping, and periodically during the casting process touch the tips of the alignment pins with a beeswax lube pencil (described in another chapter) to assure they are properly lubricated.

You should also periodically check the inner faces of the mould blocks to make certain a fleck of bullet metal is not adhering to one of the blocks thus keeping it from fully closing. Such flecks can usually be removed with a hard, flat lead carpenter's pencil filed down to a chisel edge.

As stated earlier, the premium cast bullet will have a sharp edge all around its base. If the edge is rounded, it means the bullet is not completely filled out, and while a bullet with a rounded edge might look good, you do not know for certain the radius of that rounded edge is the same from bullet to bullet, or even if it is the same all the way around on any given bullet. Discard any bullet that does not have a sharp edge all the way around its base. Such bullets are not of premium quality.

When the temperature of the melt and mould are correct and you are still getting a rounded edge on the base and sometimes on the driving bands, the cause can usually be traced to a too-tight cutoff plate. If the cutoff plate has been ground perfectly flat and the tops of the mould blocks are perfectly smooth – as they should be – a tight or even moderately loose cutoff plate will sometimes prevent air from escaping from the rear of the mould as the bullet metal is poured into the cavity.

While many excellent bullet casters like to have the cutoff plate just loose enough to swing of its own weight, bullet moulds are individualistic in their casting characteristics. If you have a mould that is giving you fits with rounded edges on the driving bands or around the base, loosen the cutoff plate until it swings freely. Then when you pour a bullet, whether it be with a ladle or bottom pour valve and spout, do not put any extra pressure on top of the cutoff plate. Merely allow the weight of the ladle to rest against the cutoff plate or merely bring the cutoff plate in light contact with the bottom pour spout, whichever the case may be.

If you use enough different bullet moulds, you will find the tightness or looseness of the cutoff plate is usually characteristic to an individual mould. That is, some moulds will demand a loose cutoff plate while others deliver superb bullets with a fairly snug cutoff plate. My own recommendation is that you loosen the cutoff plate as much as is possible and still get good bullets. This not only helps make certain the bullets are well filled out, but it also helps prevent galling between the cutoff plate and the top of the mould blocks around the pivot screw.

Another condition that will cause poor fill-out is a partially clogged ladle spout or bottom pour spout. Lead oxide collects in these holes and restricts the flow of bullet metal. I always use a Lyman dipper for my bullet casting and periodically drill out the nozzle with a 0.196 inch diameter twist drill. This is of special help when casting heavy bullets as it gets the molten metal into the mould cavity with no delay and little cooling effect resulting in a well filled out bullet.

The base of the bullet is the most critical part of the bullet. It must be perfectly flat all the way across with no protrusion or cavitation at the cutoff. Such protrusions or cavitations are seldom, if ever, located in the center of the base and thus cause an unbalanced condition on the most critical part of the bullet.

Look at the base of a bullet as the bullet travels away from you and visualize an off-balanced flywheel spinning on its axis. Being heavier on one side than on the other, the spinning bullet base soon begins to wobble and thus carry the entire bullet off its intended line of flight. The farther the bullet travels, the worse it wobbles and the farther off course it drifts.

You cannot be too fussy about the bases of your bullets. To be of premium quality, the bullet base must be perfect.

The most perfect bullet base you can get comes from a nose-pour mould. These are glass smooth with a sharp edge all the way around, and with a nose-pour mould there is never a void in the base. Further, from my limited knowledge of metallurgy, it is my understanding that the bullet metal in the base of a bullet from a nose-pour mould is denser than that from a base-pour mould. How much of an advantage this is to the shooter is a bit beyond me at this time, but a denser base would tend to shift the center of gravity closer to the base of the bullet.

Using the softer alloys such as 30-to-1 or 20-to-1 lead and tin makes it easier to cast a good bullet base than when using a harder alloy containing antimony. Although both tin and antimony are used as hardening agents in bullet metal, antimony is far more brittle than tin and has a tendency to tear out of the bullet base rather than cut cleanly. This is especially true if you try to cut the sprue while it is too hot.

I cannot emphasize too strongly the importance of taking the time and making the effort to cast premium bullets. To do it any other way is a waste of time, powder and primers.

One shooter I knew who was having problems getting a good score with a new rifle came to the range one day with a box of sample bullets he had cast. I had the opportunity to look at these bullets and saw there was a large cavity in the base cutoff area of each bullet, and this cavity was seldom centered. There was no way these bullets could be expected to shoot accurately enough at the longer silhouette ranges to obtain a good score!

Casting premium bullets is not a goal of quantity but one of quality!

3

Should I Segregate My Bullets by Weight?

In the sixth revised edition of *Wingate's Rifle Practice* by Gen. George W. Wingate, the general wrote the following:

"It has been stated that one grain in a charge of 80 grains of powder will make a difference of 9 inches at 1000 yards. At the same time, a difference of that amount, or of five grains, in the ball makes no practical difference. A larger variation, however, is fatal to good shooting."

The key word here is *practical*. Does it mean an inch in point of impact at 1,000 yards? Or does it mean a fraction of an inch? Is the target an enemy soldier or is it a steel silhouette at 547 yards or a paper target at 1,000 yards? For the two different types of shooting – combat or competition – the word *practical* could have a vast difference in meaning.

Many competitive black powder cartridge rifle shooters segregate their bullets by weight, weighing each bullet and segregating it into a group within certain parameters. For example, if they were casting bullets that averaged 500 grains, they might segregate them into four groups: 499.0 to 499.4 grains, 499.5 to 499.9 grains, 500.0 to 500.4 grains and 500.5 to 501.1 grains. All bullets that did not fall into one of these weight categories would be discarded for competitive shooting and used for practice or go back into the melting pot.

If you are casting premium bullets to begin with, this type of segregation might well get you that one extra silhouette target that wins the match or the shootoff, or it might get you that one extra shot into the X-ring at 1,000 yards, that extra edge that wins the match. It certainly is not time wasted.

Conversely, if you are trying to obtain premium bullets through sorting and segregation instead of by the casting process itself, you are wasting time segregating bullets by weight. For example, let us say you are segregating your bullets by weight as in the example two paragraphs back. If you are not using the casting techniques to produce premium bullets to begin with, how can you be certain the bullets weighing 499.0 grains do not have a void in them that causes the lighter weight? Or that the bullets weighing 501.0 grains are not larger in diameter because the mould blocks were not properly seated together?

I believe a difference in the diameter of the bullets involved would be equally, if not more, detrimental to accuracy than the 0.4 of 1 percent variation in weight. Certainly a void in the back end of a bullet is more detrimental to accuracy than the slight variation in weight, especially if the void is slightly off-center as they usually are!

If you are going to weigh each bullet and then segregate and load it by weight, you must start the process by casting premium bullets. That is, you cast your bullets within parameters of not more than plus or minus a single grain so when you weigh and segregate them, you do not have any culls to be used for practice or to be thrown back into the pot – or at least very few culls. You do this primarily by maintaining a constant temperature of the melt within plus or minus 10 degrees, 15 at the most. And you further promote premium bullets by establishing a casting rhythm that remains the same for each bullet and thus helps keep the mould temperature at a certain level. You cannot do sloppy casting and then plan on getting premium bullets by the segregation process.

As for the advantages of sorting your bullets by weight, I believe you will find most top competitive shooters follow this procedure. While one cannot argue with success, I have always striven to cast my bullets within the parameter of plus or minus one grain, sometimes closer. Beyond that, I do not weigh each bullet.

4

Safety While Casting

Most of us who have been casting bullets for the last 30 to 50 years have at one time or another been lucky. I say this because when I started casting in 1949, the whole operation was looked upon as something akin to a black art resurrected by a few shooters who were unwilling to accept certain technological advances in the shooting game and insisted on making their own. Old-time companies such as Lyman and Belding & Mull humored us with moulds, melting pots and lubricators, but beyond that we learned by doing, and some of the things we did through pure ignorance were downright dangerous.

My most memorable experience involved dumping a scoopful of salvaged backstop metal into a large cast iron kettle full of molten alloy. The resultant explosion plastered the ceiling with globs of bullet metal. I had scabs of it clinging to my neck, drops of it clotting my hair and bits and pieces on my forearms. Fortunately, both eyes were still intact though I wore no protective eyewear of any sort.

You cannot always depend upon Providence to watch over kids, fools and drunkards. Once in awhile you have to depend upon your own good common sense and better judgment in order to stay out of trouble.

Casting bullets can be downright dangerous. Even if we discount the problems of lead fumes – which we cannot – one can get severely burned by not paying attention to what goes into the lead pot or through sheer carelessness in the handling of ladles filled with molten metal. In the case of my explosion, most of the backstop metal I retrieved that long-ago day was in the form of jacketed bullets salvaged from an old, rotten chestnut stump. I distinctly recall that some of those jackets had peeled back like a banana, and that bits of rotten wood were compacted under the jagged peels of jacket metal.

When the small amount of moisture in those bits of wood was suddenly immersed in molten metal at close to 850 degrees Fahrenheit, it instantly became superheated steam and, like rifle powder, expanded in volume by several hundred times in a fraction of a second. I was lucky that day.

The situation for the present-day caster of premium bullets is probably more critical than it was at the time of my accident. Most of us who are casting premium bullets are competitive shooters and, of necessity, cast large quantities of bullets. This means we usually have a 20- to 30-pound capacity melting pot along with a heat source that will keep the pot up to temperatures as high as 850 F. or even higher. That is a lot of molten metal at extremely high temperature to be fooling around with.

I never add ingots or chunks of alloy to a pot partially filled with molten metal if I can avoid it. I put all the metal possible into the pot before turning on the heat, so if there is some hidden moisture, it has a chance to heat up and evaporate before the metal is melted. I also let the dipper warm up in the pot along with the bullet metal, so when the alloy is up to temperature, the dipper is also hot and has no moisture on it.

Regardless of how careful we are, the chance always exists for an explosion. If you are casting outdoors, there is always the possibility of a bug or moth flying into the pot. Or if you are casting in a building with a low metal roof, there is always the probability of dripping condensate on the bottom side of the roof. This condensate often does not appear until after you start the heat source and create a temperature differential between the bottom of the roof above the melting pot and on top of the roof. The minute you have a temperature differential, moisture in the air will condense on the bottom side of the metal roof.

To protect yourself from molten metal explosions, always wear protective eyewear, a soft cloth hat, long sleeved shirt with the sleeves rolled down, closed top shoes and natural fabric or leather gloves.

To protect yourself from lead poisoning, always do your casting in a well-ventilated area and always wash your hands thoroughly after handling lead alloy or cast bullets. Most lead poisoning is due to ingesting traces of lead, as lead is relatively soft and easily rubs off onto your hands. It is just as easily transferred from your hands to a sandwich, cookie or whatever snack food you choose to munch on.

In addition to a well-ventilated area, I like to set a small 6-inch fan about 4 feet from the melting pot with the stream of air directed across the top of the pot and away from me.

Finally, the casting area is no place to have children running around. If a Chip-off-the-Old-Block wants to watch Dad cast bullets, have the Chip watch from a stool or chair a short distance away. And when the Chip's span of attention has run its course, have the Chip leave the area. Don't let him or her stand at your elbow where they might breathe lead fumes or be struck with molten spatter.

Now sooner or later you will always have the adult visitor who doesn't know siccum about casting bullets and wants to observe firsthand just how this black art is accomplished. He might be a good friend or he might be one of those who barely tolerates your activity but feels he must show some interest in it. While you have to be the judge of this and determine in your own mind just how far you should go in giving safety precautions, be advised that the first thing this person will do is pick up a bullet just as you dump it from the mould. The second thing he will do is drop the bullet faster than he picked it up. That's about the time I look at him and grin and say, "I was going to tell you about that, but you moved too fast for me!"

Be assured your visitor will not pick up a second bullet!

5

Maintaining the Bullet Mould While Casting

The bullet mould is a precision instrument and should be treated as such. One small indentation on the edge of the cavity or around the rim of the base is sufficient to ruin the mould as far as mid- to long-range accuracy is concerned, and perhaps sufficient to ruin the mould for all casting purposes.

I can think of no chore more frustrating than trying to free a bullet from the half-cavity of one mould block. Yet that is precisely what you ask for when you store loose mould blocks in a container that lets them rattle around like peanuts in a wash tub. It is precisely what you ask for during the casting process when you slam the blocks together without first aligning the alignment pins with their respective alignment holes. It is precisely what you ask for during the casting process when your mould is carelessly moved about the work bench or placed on or about the melting apparatus in such a manner that the open blocks can strike other metal objects. Once so damaged, the mould can never be repaired.

Good bullet mould maintenance starts before the casting process. Whether you are using a brand new set of mould blocks or a much used bullet mould, the first thing on the agenda is to clean the blocks of all oil, WD-40 or other preservatives. Since I always spray my moulds with WD-40 after they have cooled following a casting session, just before the next casting session I wipe the blocks dry with a clean paper towel. The blocks are then thoroughly sprayed with Prestone Starting Fluid to remove all traces of oil, etc. This is an ethyl ether formula that does a good job of cleaning and quickly evaporates afterward.

Prestone Starting Fluid is also highly flammable. Use it outdoors. Do not use it in your casting area or any place where it could be ignited by an open flame or spark.

Once the mould is thoroughly cleaned, it should be preheated to the point where it will melt beeswax that is touched to the mould but not so hot that it will cast good bullets. I say this because I believe more bullet moulds are warped during the preheating process than during the casting process.

Once the mould is preheated enough to melt beeswax, take your beeswax pencil and lightly touch the top surface of the cutoff plate in one or two spots and on the lip of the bevel around the pour hole. You do not need nor want too much beeswax on the cutoff plate, just enough so the sprue will not stick to the cutoff plate once it has been cut free from the bullet.

Drawing the beeswax pencil across the bottom of the mould blocks and along the sides of the mould blocks will help prevent spatter from adhering to these surfaces. Open the blocks and touch the tips of the alignment pins with the beeswax pencil. This lubricates the pins and helps prevent wear around the edges of the matching alignment pin holes.

Finally lubricate the cutoff plate pivot screw along the edge of the plate adjacent to the screw. This will allow a small amount of beeswax to seep between the top surface of the mould blocks and the bottom surface of the cutoff plate, thus lubricating a very critical area.

During the past six months I have seen two brand-new Lyman bullet moulds ruined due to lack of lubrication on the pivot screw. These were both high-quality silhouette moulds, one for the superb .40-caliber Snover bullet and the other for Lyman's new version of the long-proven 457132 Postell bullet.

How did it happen? Easy! The pivot screw holding the cutoff plate was tightened just enough so the cutoff plate would not swing freely by its own weight, and at the same time did not bind or show excessive drag across the top of the mould. Nothing wrong here.

But when you have a steel cutoff plate sliding or rotating upon a cast iron block, unless you keep the surfaces of the two members lubricated, there is a strong tendency for the metal in the cast iron block to adhere to the steel cutoff plate and vice versa. They call this galling, and once it starts, the buildup of metal on the two parts progresses rapidly.

Within a short time the cast iron around the pivot screw is worn away so the cutoff plate seems loose. You tighten the pivot screw and more metal wears away until suddenly you notice the base of the bullet has a flashing around its rim. Now, the more you tighten the pivot screw, the more its far edge lifts above the mould blocks and the greater is the flashing. Yet if you loosen the pivot screw so the cutoff plate lies flat, it flops around during the casting process, and you still often have a flashing. In short, the mould is ruined.

When you are casting bullets, you must keep the area around the pivot screw lubricated. The instant you feel a tightness or increased drag on the cutoff plate, lubricate the pivot joint surfaces by touching the edge of the cutoff plate adjacent to the pivot screw with the tip of the beeswax lube pencil.

Very little lubricant is needed to keep the pivot joint free of galling. In fact, if you get too much melted beeswax here, it will spread across the tops of the mould blocks and into the mould cavity, covering the bases of the bullets with tiny bubbles and the sides of the bullets with wrinkles. Your bullets will continue to be this way until the beeswax is either burned away by continuous casting or is wiped away with a soft clean cloth kept handy for just such occasions.

In addition to periodic lubrication of the pivot screw, you should also touch the tips of the alignment pins with the beeswax pencil. Keeping these pins lubricated helps prevent wear around the opposing alignment pinholes and thus promotes better alignment of the mould blocks. It also helps the blocks to seat tighter against one another in those cases where the pins are a bit sticky in the alignment holes. All this better alignment and more uniform closing of the blocks contributes toward more uniform, better balanced premium bullets.

Once in awhile you will have a bullet mould on which the sprue has a tendency to stick to the beveled cutoff cavity on the cutoff plate. If you have a large puddle of bullet metal making up the sprue, it can usually be tapped loose by striking the joint of the mould handles while holding the blocks bottom side up. But when the sprue fills only the beveled cutoff cavity, it is so light in weight it sticks to the cutoff plate like beggar's lice.

Whenever the sprue shows any indication of sticking, merely touch the lips of the bevel around the pour hole with the tip of the beeswax pencil. Again, you do not need nor want an excessive amount of beeswax here, just enough to make a small wet spot on the lip of the bevel.

In the event a flake of bullet metal adheres to the tops of the mould blocks or the bottom of the cutoff plate, do not try to scrape it off with a screwdriver blade or the sharp edge of your jackknife. For cases like this, you should always have within reach a Dixon red and black 997-H carpenter's pencil having a 3/16 inch wide hard pencil lead. With about 1/4 inch of the lead exposed, work the pencil back and forth on a fine cut file until you have a sharp wood-chisel edge on the pencil lead. Use this to cut flakes of bullet metal from the mould blocks or cutoff plate. Since the pencil lead is softer than the mould blocks, it will not damage them as will the sharp-edged screwdriver or knife blade.

While casting bullets, one should always keep a close eye for small flakes of bullet metal on the mould blocks or any buildup of bullet metal smear on the bottom of the cutoff plate. Flakes of lead on the inner faces of the mould blocks will result in a heavier, larger diameter bullet. Although the difference may be small, it is still there; it is these small differences that separate premium cast bullets from the run-of-the-mill cast bullets.

Any buildup of bullet metal smear on the bottom of

the cutoff plate or tops of the mould blocks is usually the result of cutting the sprue too soon. Give the sprue plenty of time to harden.

Since even the hard pencil lead in these carpenter's pencils is relatively soft when compared to the metal in the mould blocks, it will pay to keep two pencils sharpened and ready to use at the beginning of each casting session.

One final note. As I have said before, a bullet mould is a precision instrument made by some of the finest craftsmen the world has ever seen. When you are casting bullets, do not beat your mould to death by using a club or mallet to cut off the sprue. Wear a pair of gloves and swing the cutoff plate by using the heel of your hand against the projection on the cutoff plate and exerting a slight downward pressure against the mould blocks at the same time. This will help give you a clean cut base that is square with the axis of the bullet.

6

Feeding the Riser

In the old iron foundries, they often used the expression "feed the riser" when pouring the molten iron into the mould. These were sand moulds, and the riser was the hole extending down through the top of the mould into which they poured the melt. The term "feed the riser" meant to keep the riser full of molten metal to compensate for shrinkage and to reduce porosity as the metal cooled within the mould. Without feeding the riser, sand castings were apt to be porous and have voids of various sizes.

Casting the premium bullet is no different. Regardless of whether you are using a dipper or a bottom pour valve, you have to "feed the riser" in order to get the most dense bullet possible with as few voids as possible. You have to "feed the riser" if you want the base of your bullet to have a sharp edge all the way around instead of a radius or rounded edge.

Feeding the riser as related to casting bullets means to

hold the dipper against the mould for several seconds after the mould is filled or to maintain a steady stream of bullet metal from the bottom pour melting pot for several seconds after the mould has been filled. In both cases the object is to keep feeding bullet metal into the mould as the metal already inside begins to cool and shrink. This will reduce the chance of a void within the bullet and will assure a good, sharp edge around the base of the bullet.

Some bullet casters who use a dipper pour the metal into the mould from a height of ½ to 1 inch instead of holding the spout of the dipper against the mould to take advantage of the weight of the metal within the dipper. In so doing, their bullets often have a slight radius around the edge of the base instead of a clean sharp edge. This is because the Lyman dipper and other similar dippers do not have the capacity for "pouring" that you get from a bottom-pour melting pot.

If you are going to pour your bullets from a dipper, then you need a Rowell bottom pour ladle produced and marketed by Advance Car Mover Co., Rowell Division, PO Box 1, 240 N. Depot St., Juneau WI 53039. Its No. 2 ladle has a two-pound capacity, more than enough to accommodate a single- or double-cavity mould. Although I believe a No. 1 one-pound capacity ladle would be sufficient and easier to use, I opted for the two pounder and have never tried the smaller one.

The Rowell ladle is designed so that molten metal is poured from the bottom of the ladle instead of the top, thus helping prevent the inclusion of oxides or other impurities sometimes found on the surface of the melt.

Although I have used both styles of dippers – the Lyman and the Rowell – and get equally superb bullets from each, I prefer to hold the nozzle of the dipper against the mould rather than to pour the melt from above the mould.

The next logical question is how long should I feed the riser? Or how long should I hold the dipper against the mould? The answer is that you feed the riser long enough so when the flow of bullet metal is shut off or the dipper is removed from the mould and you have left a generous puddle of melt for a sprue, there is little or no indication of the puddle being sucked into the mould as the metal starts to cool. If an excessive amount of puddle is being sucked into the mould, you will see a prominent dimple on the surface of the puddle right over the hole in the cutoff plate.

This is always an indication that you stopped feeding the riser too soon; usually there is a void within the base of the bullet. You can check this by grasping the nose of the bullet in a pair of pliers and melting away the base with a propane torch. As the base melts, the void will be exposed.

As to how long, in units of time, you should feed the riser, this could be affected by the composition of the alloy and the temperature of the mould and melt. Although five seconds is usually sufficient, the amount of time should be established by whether or not the sprue puddle is dimpled. In my opinion, if there is an apparent dimple in the center of the sprue puddle after it has cooled, you have not fed the riser long enough. You may have a good bullet or you may not.

As an experiment, on December 27, 1997, I cast 49 bullets, laid them out on my casting board in the order cast and weighed them on my Lyman LE-1000 electronic powder scale. My casting temperature was held between 820 degrees and 830 degrees Fahrenheit, and the mould used was the Ballard-Postell cut on Lyman blocks.

The first 15 bullets were cast at my normal rate in order to make certain the mould was at the proper casting temperature. For the next 12 bullets, the dipper was held against the mould for the count of five (5 seconds); the next 11 bullets for the count of 10 (10 seconds), and on the final 11 bullets I held the dipper against the mould until all the metal had run out. Sometimes this was more than 15 seconds and sometimes it was closer to 10 seconds. The variation in bullet weight is tabulated as follows:

Warm-up	5 Seconds	10 Seconds	Run Out
547.4	547.3	547.1	547.4
548.1	547.3	547.2	547.3
547.5	547.5	547.0	547.0
548.1	547.2	546.9	546.3
547.7	547.4	547.3	546.3
547.4	546.8	547.2	546.5
547.9	547.0	547.1	546.5
547.5	546.3	546.9	547.1
547.4	546.7	547.1	547.1
547.9	546.9	546.9	546.4
547.5	546.4	547.0	546.5
547.7	546.7		
547.7			
547.5			
547.4			

	Average Weight (*grains*)	Weight Spread (*grains*)
Warm-up	547.646	0.7
5 Seconds	546.958	1.2
10 Seconds	547.063	0.4
Run Out	546.763	1.0
Overall	547.107	1.8

While the number of bullets cast within each category is far too limited to establish any definite conclusions, there are a couple fairly good indicators. First, although the mould was preheated prior to any casting, it is readily apparent the bullets cast during the warm-up period were heavier than those cast later. And second, it would appear that feeding the riser for 10 seconds provided the most uniform bullets as far as weight is concerned.

To see if this trend would prove itself throughout an extended casting session, on May 12, 1998, I cast 216 bullets from the same mould at a casting temperature of 760 degrees Fahrenheit. The lightest bullet of the lot weighed 546.8 grains while the heaviest weighed 548.1 grains, an extreme spread of only 1.3 grains or a bit more than one-fifth of one percent of the entire bullet weight. The first bullet cast after the mould had been brought up to casting temperature weighed 547.8 grains, while the last bullet cast weighed 546.9 grains.

Feeding the riser for a period of 10 seconds on each bullet cast slowed my casting rate by 7 bullets per hour, to a total of 78 per hour. But of the 216 bullets cast, only four weighed less than 547.0 grains and only 14 weighed more than 547.9 grains. In short, of all the bullets cast, 198 of them weighed within plus or minus 0.5 grain.

Casting at the lower temperature of 760 versus the 820 degrees of the earlier test had its advantages. The sprue hardened faster, and the base of each bullet was perfectly flat with little or no hint of cavitation. Further, I was able to start casting 15 minutes earlier than usual because I didn't have to wait for the melt to heat up the extra 60 degrees.

All in all, casting at the lower temperature and feeding

the riser for 10 seconds on each bullet gave me better quality bullets with only a slight loss in casting rate, a negligible price to pay for the premium cast bullet!

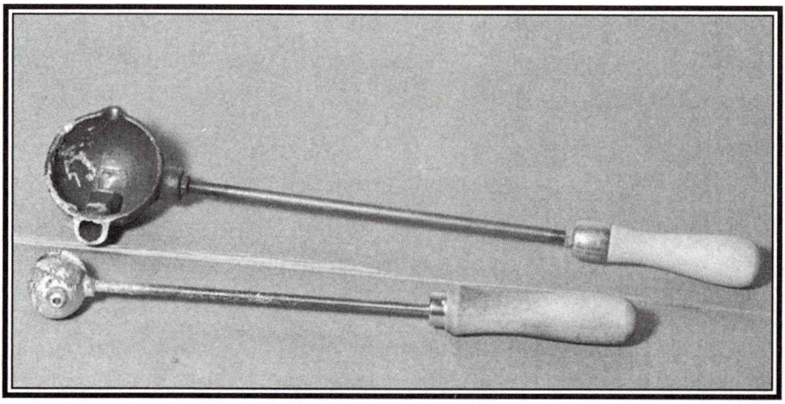

Two entirely different casting dippers: on the bottom is the Lyman dipper purchased in 1949. I still use that same dipper today, though with a new handle. The upper dipper is the bottom-pour No. 2 Rowell two-pound capacity dipper. I would recommend that anyone contemplating this style dipper purchase the No. 1 one-pound capacity model.

7

Constant Temperature and Heat Source

If there is any one factor vital to casting the premium bullet, it is maintaining a constant temperature of the melt and mould during the casting process. The weight of a .45-caliber bullet whose average is 550 grains can vary as much as 2 grains below average if the mould is not up to regular casting temperature to 2 grains above average if the mould were preheated and allowed to get too hot. This is a differential of 4 grains!

If you are casting at a temperature of 820 degrees Fahrenheit and starting with a preheated mould that is casting good bullets but is not yet up to casting temperature, your first 8 to 12 bullets will run a bit lighter than the norm. And if you take a break during the casting process to answer the telephone, the first 8 to 12 bullets you cast after resuming operations will most likely run a bit lighter due to the fact the mould cooled while you were talking on the telephone.

I cannot emphasize too strongly that when casting

bullets you must always maintain a constant temperature of the melt and mould in order to get the least variation in bullet weight. How small a variation is possible? Just about as small as you are willing to take the time and make the effort to get. For myself, I try to cast within plus or minus one grain or less for an entire casting session. Others have written that they can cast to a closer tolerance than that, and from my own experience, I believe them.

There is no magic in maintaining a constant temperature of the melt and mould. It is strictly a matter of equipment and casting rhythm: the equipment being a heat source that can be precisely regulated and an RCBS or similar lead thermometer and the rhythm being a repetition of the same motions during the same time span for each bullet cast. It is as simple as that.

You cannot talk about maintaining a constant temperature and not talk about a heat source, and you cannot talk about a heat source without some discussion or thought about the bullets you cast and how you cast them.

Most shooters just getting into shooting cast bullets in the black powder cartridge rifle have already had some experience at the casting process, and most already have a thermostat regulated electric melting pot of about 20 pounds capacity equipped with a bottom pour valve and spout. This is fine for casting lightweight pistol bullets or for smallbore rifle bullets up to 220 or 250 grains. But when you start getting serious about shooting the black powder cartridge rifle, you start thinking about .40 caliber and larger with bullets from 400 grains up through 550 grains for the .45s, and 650 or 750 grains for the Big Fifty. Casting quantities of premium bullets for these larger calibers demands a whole new look at the casting process.

First, consider capacity. It takes one pound of bullet metal to make 12.7 bullets weighing 550 grains, or about 254 bullets from 20 pounds of bullet metal. While this may sound good, it isn't. You want to be able to cast your bullets and at the same time have sufficient pot capacity so you don't have to add more bullet metal during the casting process. Adding more metal lowers the temperature of the melt and the mould resulting in a drastic change in bullet weight until everything is up to temperature again.

Second, do you use a bottom pour spout or do you use a dipper to get bullet metal into the mould? Most shooters who have been at the game for any length of time use one of the cast iron ladles offered by Lyman or RCBS for pouring metal into the bullet mould. Over the years they have found that when casting large diameter heavy bullets, they get a much better bullet from the ladle than from the bottom pour spout.

Using a ladle for casting premium bullets also presents its problems. If you use a ladle, the melting pot must be large enough so the ladle never rubs against the sides or bottom of the pot during the dipping process. Rubbing the ladle against the sides or bottom of the pot usually stirs up some grit that gets imbedded on the base or exterior of the bullet – a condition not conducive to premium bullets.

What all this adds up to is that if you are going to cast quantities of large caliber heavyweight bullets of premium quality for the black powder cartridge rifle, you should start thinking about a 35 or 40 pound capacity melting pot and a heat source that will easily keep that much melt at a temperature of 850 degrees Fahrenheit.

As I write this, I am in my 50th year of casting bullets. As such, I have melted my bullet metal on all sorts of appliances from my wife's kerosene stove (not recommended!) to a home-made coal furnace, gasoline and propane plumbers' furnaces, a Lyman electric bottom pour furnace, two different Coleman stoves and two different propane outdoor cookers manufactured by Metal Fusion, Inc. (712 St. George Avenue, Jefferson LA 70121). Of all I have tried, the latter two are by far the best as well as the most economical to purchase and the most economical to use.

In 1995 I purchased a Model 99PK outdoor cooker on which I used a 2-quart cast iron sauce pan as a melting pot. This is a 36,000 BTU cast iron burner that comes equipped with a low pressure regulator, three foot length of hose with fittings and a needle valve. To use with it, I also had to purchase a 20-pound propane cylinder.

This is a superb unit that is easily regulated to plus or minus 10 degrees Fahrenheit. I have cast thousands of bullets with it over the past four-and-a-half years and, in general, have been well satisfied with its performance.

As for the downside of the Model 99PK, its heat output is marginal for the job we are trying to do. Without a heat shield around the melting pot to keep the heat concentrated, you will have a tough time getting the temperature of the melt up to 820 degrees F. With a heat shield – a circular piece of heavy gauge sheet steel 3 inches high and 7½ inches in diameter – you can keep 25 pounds of bullet metal at 850 degrees F. without problem. This is about ¾ the full capacity of a 2-quart cast iron sauce pan.

Because of the marginal capabilities of the Model 99PK, in the summer of 1999 I purchased a Model TCI outdoor cooker having a BTU output of 135,000. This is a high-pressure burner that comes equipped with the proper hose and fittings and a high-pressure regulator. It does not have a needle valve like the 99PK. You adjust the flame by adjusting the high-pressure regulator.

With this outfit you can be casting bullets at 850 degrees a half hour after lighting the burner, this with the pot brimming full with 35 pounds of bullet metal. The downside of this outfit is the difficulty in maintaining a constant temperature. The slightest movement of the adjustment screw on the pressure regulator sends the heat in one direction or the other requiring that you keep an eye on the thermometer at all times. Later, I purchased another hose with fittings that included a needle valve, and this helped alleviate the problem.

Another superb high-pressure outdoor cooker is the Model HD-26 manufactured and sold by the Marrone Company (465 Albert Street, Macon GA 31216). This is a rugged 170,000 BTU propane burner on which the flame is regulated by a needle valve instead of an adjustable high-pressure regulator. It was from the Marrone Company I purchased a preassembled hose with fittings and needle valve to use on my own TCI burner.

I should emphasize that these burners are advertised and marketed as outdoor cookers. If you choose to use one inside for bullet casting, leave the propane tank outside and have a special length of high-pressure hose made up to connect the cooker with the cylinder. *Never use a propane gas cylinder inside a building, and always remember to shut off the valve at the cylinder as well as at the cooker.*

WARNING:** **It is very easy when using one of these high-pressure outdoor cookers to get the temperature of the melt well above 850 degrees and even over 900 degrees Fahrenheit. However, at these high temperatures, the bullet metal will start to vaporize, and you may be subjected to lead poisoning by breathing the fumes. Do not heat your bullet metal any hotter than 850 degrees F.

For those who do not want or cannot use a propane burner, a good electric melting pot of at least 20 pounds capacity with thermostat control should serve you well. Lyman offers its excellent Mag 20 electric furnace with a capacity of 20 pounds, and RCBS offers its bottom pour Pro-Melt having a capacity of 22 pounds. For the bullet caster who uses the dipper for running bullets, Lyman also offers a 20-pound capacity Magdipper Casting Furnace.

You should be aware when using any furnace with a thermostat control that as the thermostat ages the heat output at any given setting may decrease. That is, if you set the thermostat at 800 degrees F. when it is new, a year later the same setting might give you only 775 F.

Thus, regardless of whether you use a gas burner or an electric melting pot, you should also have an RCBS or Lyman lead thermometer in the melt at all times during a casting session. This is critical if you are going to cast premium bullets. You do not want the temperature of the melt to fluctuate more than 10 degrees in either direction, less if you can control it.

Finding a means of keeping the thermometer in the

melt at all times can be a problem. A clip on the stem of the thermometer is usually worthless. Even if the side of the lead pot is thin enough to accommodate the clip, the high temperature of the pot quickly eliminates the spring temper of the clip so it no longer holds the thermometer securely.

A little bit of ingenuity will usually solve the problem. What is needed is a short piece of tubing attached to a C-clamp or to some kind of a stand with an arm on it to overhang the lead pot. The tubing should have a large enough inside diameter to accommodate the stem of the thermometer with room to spare.

If the lead pot is shaped correctly, a C-clamp with the tubing attached can be clamped to a convenient spot along the rim of the pot. If for some reason this cannot be done, a stand of some sort having an arm with the tubing attached can be used to overhang the rim of the pot and support the thermometer.

The stand I use was once used by a machinist to support a dial indicator while trying to center a piece in a lathe chuck. This works well for me and happened to be what I had on hand at the time. Lacking this ready-made stand, I would have taken a piece of stiff wire or rod, bent it into the shape desired and anchored the base end in a small can filled with melted lead. This would have served as well as anything else including the dial indicator stand.

Once you have the proper equipment for casting a premium bullet, you must establish a timed rhythm to your casting process. That is, you must do everything the same way and in the same amount of time for every bullet you cast. If you are dipping your bullets, you must fill the dipper or ladle the same amount each time and hold it against the mould for the same duration of time for each bullet. You must let the sprue harden for the same amount of time for each bullet and you must refill the mould at precise intervals during the casting process.

All of this "same time" business is to keep the mould at a constant temperature, thus helping to assure a consistency in the weight of each bullet. Some of the very dedicated accuracy-minded cast bullet shooters even go so far as to mount a thermometer on the bullet mould in order to assure a constant temperature and cast bullets that weigh within a couple *tenths* of a grain of each other.

This is precision casting, and it can be done! While I do not go so far as to mount a thermometer on the mould blocks, I am fussy about maintaining a constant rhythm to my bullet casting and a specific temperature of the melt. Doing these two things will give you bullets within plus or minus one grain.

If you wish to sort your bullets by weight after that, fine. Beyond checking the first few bullets I cast against those I finish with and a few in between, I do not segregate my bullets by weight. I do label each lot of bullets cast during a single session by date and average weight of the bullets within that lot. And personally, I believe this is close enough, especially when dealing with bullets of 500 grains or over.

After all this talk about maintaining a constant temperature of the mould and melt, the next logical question is what is the best casting temperature? Very simply put, the best temperature for casting bullets is the *lowest* temperature at which you can get a good bullet at every opening of the mould. This can vary all over the lot depending upon the alloy, style of bullet, material from which the mould is made and whether the mould is base pour or nose pour. But regardless of what you are working with, go for the lowest temperature that gives good bullets.

There are several reasons for this, the first being that it is more economical to cast at a low temperature than a high one. Second, there will be less fumes from the bullet metal to create a health hazard; and third, a low casting temperature lets the sprue harden a bit faster and helps promote a more perfect base on the bullet with no hint of cavitation at the cutoff.

Any cavitation on the base of the bullet, regardless of how slight, is detrimental to accuracy. The cavitation is never symmetrical, and it is almost never in the center of the base. Thus, when you shoot a cast bullet that has a slight cavity in the base, that bullet is unbalanced before it ever leaves the chamber. It does not improve on its journey down the barrel; and upon exiting the muzzle, the base of the bullet will start to orbit around its theoretical axis – wobble. The farther the bullet travels, the worse it gets, and instead of hitting that ram target at 547 yards, the bullet is either over the back or under the belly, anywhere except where the sights lay at the time of firing.

Cavitation on the base of the bullet is partially caused by cutting the sprue while it is too hot. Although the outside surface of the sprue may appear solidified, and indeed is solidified, the inner core of the bullet is still very hot and very soft. When you swing the cutoff plate to shear off the sprue, a small portion of soft metal is pulled from the base of the bullet.

Casting at a low temperature helps prevent this. For

example, if you are in the habit of casting at 820 degrees Fahrenheit when you can get good bullets by casting at 760 degrees Fahrenheit, then you have a sprue that is 60 degrees F. hotter than need be. It takes time to cool the sprue that extra 60 degrees, time that most of us would rather spend casting the next bullet. And because most of us are impatient, we cut the hot sprue prematurely and go on with the next bullet despite the fact the bullet we just dumped from the mould was less than perfect.

Bullets that are less than perfect are not of premium quality.

This is the Lyman Mag 20 melting pot with thermostat control for the bullet caster who uses a dipper. If you want to cast premium bullets with an electric furnace, this is an excellent choice. (*Courtesy of Lyman Products*)

The Lyman Mag 20 electric melting pot has a valve-controlled, bottom-pour spout. Though I much prefer to dip my bullets, many bullet casters prefer the electric furnace with a bottom-pour spout. (*Courtesy of Lyman Products*)

The 99PK burner produces 36,000 BTU. Shown here with a heat shield at right and an RCBS lead thermometer and stand at left, this makes an excellent unit for casting bullets. This is a low-pressure burner and is easily controlled.

The TCI burner has a 2-quart cast iron sauce pan for a melting pot and an RCBS lead thermometer. This unit produces 135,000 BTU and will bring 35 pounds of bullet metal up to 850 degrees F. in about 20 minutes. It is a good heavy-duty casting outfit.

8

How to Warp a Bullet Mould!

I believe few shooters ever warp a bullet mould, but I also believe this bit of fortune is more happenstance than by intent simply because most bullet casters use an electric melting pot instead of a gas burner.

Bullet moulds seldom suffer any warpage due to casting bullets. The warpage occurs during the preheating process when the mould is overheated, and this is the reason those using the electric melting pots experience few problems of mould warpage. The mould is usually stood upright on the flat rim of the melting pot or on a small shelf designed expressly for that purpose and does not overheat during the time the bullet metal is melting.

For those using a gas or propane burner, the situation is far different. Here the mould is often laid across the top of the burner with the mould blocks an inch or so away from the flames. Or, in some cases, the mould is propped up with the mould blocks resting on the rim of the melting pot some four or five inches away from the flame. In either case, the mould blocks are vulnerable to warpage because of the extreme radiant heat of the gas flame. Believe me, I know whereof I write, and I have the warped mould blocks to prove it!

If you are using a large capacity propane burner as a heat source for casting bullets, do not expose your bullet mould to the preheating process for an extended period of time. Instead, watch the lead thermometer, and when the temperature is about 20 degrees under your casting temperature, prop the mould blocks against the rim of the melting pot for the short time it will take the melt to come up to casting temperature. Better yet, bring the mould up to casting temperature by casting and discarding the first dozen or so bullets. This is the best way to prevent mould warpage.

It doesn't require much effort to warp a mould. You can do it without half trying! And it is always one of your prize moulds that suffers the consequences.

Virtually all of us preheat our bullet moulds by placing them somewhere near the heat source so they will have reached casting temperature at about the same time as the melt. That way, very little time is wasted when the actual casting process begins. The first bullets cast are well filled out with no "cool" wrinkles.

When I first started using a propane burner, I placed the mould across the top of the grill so the blocks were an inch or two away from the flame. I did this a number of times with no apparent problem, and the mould was usually at casting temperature when I started casting.

Then one day when I cast the first bullet, I noticed it took the sprue puddle 15 or 20 seconds to set, far longer than normal. Realizing the mould was far hotter than normal, after dumping the bullet I held the blocks open for about 30 seconds to let them cool and then went on casting.

At the next casting session, instead of preheating the mould on the grill, I propped it up with the blocks resting on the rim of the cast iron pot, at least four inches away from the flame. This worked fine as the mould was just under casting temperature when I started casting, and I continued to preheat whatever mould I was using at the time in this fashion.

Somewhere along the line I noticed my prize Ballard-Postell mould was slightly warped – if closed and held to a light, a sliver of light would show between the blocks at the bottom of the cavity. However, I continued to use the mould, and the bullets shot as well as ever. As far as the warpage was concerned, I blamed that on the overall size of the blocks as they were not quite as large as RCBS or SAECO.

The next step in my education on warping bullet moulds came about in using an RCBS 45-500-FN mould that had been well oiled and stored away for several years. It had always cast excellent bullets, but this time for some reason it did not. A suggestion was made that perhaps the mould was contaminated by the oil and perhaps if it were heated in an oven at 450 degrees Fahrenheit for one hour, that would bring the oil out of the pores in the metal.

Had I followed that advice, I might have been all right. Instead I heated the mould on top of a wood heating stove for an hour and discovered I had warped it in the process. This brought back recollections of the Ballard-Postell mould and the fact that preheating the mould to too hot a temperature had caused the warpage, not the small overall size of the blocks.

With this bit of knowledge firmly implanted, I vowed to never warp another mould due to overheating during the preheat process. From then on, all moulds were preheated by propping them up with the blocks on the rim of the melting pot at least four inches away from the flame.

This worked fine for a number of months. Then came the day that in order to get the alloy correct, I had the pot brimming full of metal. This, of course, increased my melting time as well as the length of time my brand new Paul Jones Creedmoor bullet mould was propped against the rim of the pot.

When it came time to start casting bullets, I noticed the sprue remained fluid for several seconds longer than normal on the very first bullet. After the casting session was finished, I could see that once more I had warped a bullet mould – not badly, but warped nonetheless.

The upside of all this is the moulds involved continue to cast as well or maybe better than ever before with no discernible change in dimensions, weight or accuracy. There is no flashing along the parting line on the nose of the bullet – the area where the moulds are warped – and if you were not told the moulds were warped, you would not be aware of it by the appearance of the bullets cast.

The downside of the situation is I know that through my ignorance I have warped three very special bullet moulds I would rather not have warped. And you can bet I'll never warp another one during the preheat process!

With all this said, after I had warped the RCBS mould – one that was warped more than the others – I considered methods of trying to straighten the RCBS blocks. First, I removed them from the handles and took off the cutoff plate. Then, with the two blocks mated together, I clamped them in a vise for a few minutes, afterwards holding them up to a light to see if a sliver of light were still visible at the nose end of the cavity. It was, but it was also greatly reduced from what it had been.

After repeating this process two or three more times with no further improvement, I clamped the mated blocks in a vise and left them overnight. In the morning, I could no longer see the sliver of light at the nose end of the cavity until after I separated the blocks and then mated them together again. At that time I could see a fine sliver of light.

Convinced my efforts had failed, I mounted the blocks on a set of RCBS handles, preheated them by casting a few bullets and then lubricated the tips of the alignment pins with a beeswax lube pencil. Before casting another bullet, I glanced inside the cavity looking for that telltale sliver of light. It was gone! With the blocks on the handles, hot or cold, the RCBS mould blocks appear perfectly mated with no apparent warpage.

To this day I have not tried to correct the other two moulds, nor do I plan to. They still cast premium bullets that give excellent accuracy, and I am not going to take a chance of further damage. I merely tell you what I did to correct a bad situation with the RCBS mould blocks and leave the decision up to you as to whether or not you want to try to correct a mould you may have warped.

As for avoiding warpage during the preheat process,

my advice is that if you are using a propane burner, do your preheating by casting. It may take a bit longer, but it will not warp your prize bullet mould!

9

Making a Beeswax Lube Pencil

If there is any one thing that will ruin a good bullet mould and for which there is no excuse, it is lack of lubrication during the casting process. In view of the fact we are told to thoroughly clean the mould of all grease or oil prior to casting, the idea of lubrication seems like double talk. It isn't!

Harry Pope once said plain yellow beeswax was the best lubricant for a bullet mould, and for years I have followed his advice. I kept a small block of beeswax on the casting bench and two or three times during each casting session I would scrape off a small curl of beeswax with the blade of a narrow screwdriver. With this beeswax I would touch the edge of the cutoff plate adjacent to the pivot screw, and I would touch the tips of the alignment pins. Once in a great while I would very briefly touch the top of each mould block being very careful not to get so much beeswax on the mould that it went into the mould cavity.

All this worked fine except sometimes the screwdriver was not where I thought it was, and other times either I did not get a curl of beeswax when I drew the blade across it, or else I would get too large a curl of beeswax. The situation begged for something better.

My answer to the problem was to devise a stick of beeswax about the diameter of a pencil and 5 or 6 inches long. To keep it from picking up dirt and grit, it is enclosed in a paper wrapping, so one need only unwrap a short bit to expose sufficient beeswax to work with for a long time. I call this a "beeswax lube pencil."

The beeswax lube pencil is simple to make, easy to use and will last a long time on the casting bench. To make one, you will need 2 or 3 ounces of yellow beeswax, a one-cup capacity Pyrex measuring cup, a glue stick (Elmer's works fine), a 4x6-inch plain note pad and a dowel about 7 inches long and 5⁄16 inch in diameter. For the latter, I use a section of aluminum arrow shaft. A 3x5-inch note pad will work as well as a 4x6; it will just give you a shorter lube pencil.

The object of this operation is to make a paper tube closed off at one end and filled with melted beeswax. You may find a simpler way to achieve that objective, but here is how I do it.

1. Cut a 2-inch square from a sheet of note paper.
2. Place the end of the dowel in the center of this square and fold the edges of the square around the dowel. This will provide a cap with which to close off one end of your paper tube.
3. Lay a full sheet of note pad paper on the table in front of you with the long side of the sheet running

left and right. Using the glue stick, liberally coat the sheet across one end, the end at your left, with glue.

4. Lay the dowel, with the paper cap on the end to your left, along the long edge of the sheet with the paper cap flush with the glued end of the sheet.
5. While holding the edge of the sheet tightly against the dowel, roll the dowel across the sheet to form the tube. Stop rolling about ¾ inch from the far edge of the sheet and apply a liberal amount of glue along this edge.
6. Continue to roll the dowel and tube and lightly press the glued edge against the full length of the tube and around the cap.
7. Withdraw the dowel and set the tube aside for half an hour to let the glue set.
8. While the glue is setting, melt the beeswax (2 or 3 ounces) in the Pyrex measuring cup. I recommend you melt the beeswax by setting the measuring cup on a paper towel folded in half twice in a pan of water. Heat the water to just under boiling.
9. After the beeswax is melted, hold the closed end of the paper tube in the melted beeswax long enough for the wax to impregnate the paper. Usually about 30 seconds is sufficient. This is to seal the bottom end of the paper tube.
10. Withdraw the paper tube and let it set for at least half an hour so the beeswax can completely cool.
11. After the wax has cooled, stand the paper tube upright in a paper cup. Hold it in an upright position by gently packing Kleenex or similar tissue around it. Be careful not to collapse the paper tube.

12. Pour a little melted beeswax into the tube, perhaps enough to fill half an inch to an inch, and let it stand for a couple of minutes to solidify.
13. Pour in the rest of the beeswax until the tube is filled. As the wax inside the tube cools and shrinks, keep adding more wax until the tube is completely filled.

After the beeswax lube pencil has cooled and hardened, take a razor blade or utility knife and cut a short slit in the paper tube beginning at the open end of the tube. With the tip of the blade, lift the outer corner of paper from the beeswax, unwrapping it to expose about ½ inch of beeswax.

This beeswax lube pencil is the handiest thing I know of for lubricating a hot bullet mould during the casting process. I keep mine within easy grasp on a shelf right above the casting bench and use it just about once an hour while casting to keep the pivot screw and alignment pins well lubricated. Once in a great while I touch the tops of the mould blocks to apply just enough lube to prevent any galling across the tops of the blocks.

Very little beeswax is required to lubricate the mould. In fact, if you get too much wax on the mould, it will get into the cavity and cause wrinkled bullets for a dozen or so castings. Too much beeswax will also cause tiny dimples and imperfections on the base of the bullet.

If you are going to cast premium bullets, take the time to make a beeswax lube pencil, and then use it. Don't ruin a good mould for lack of lubrication.

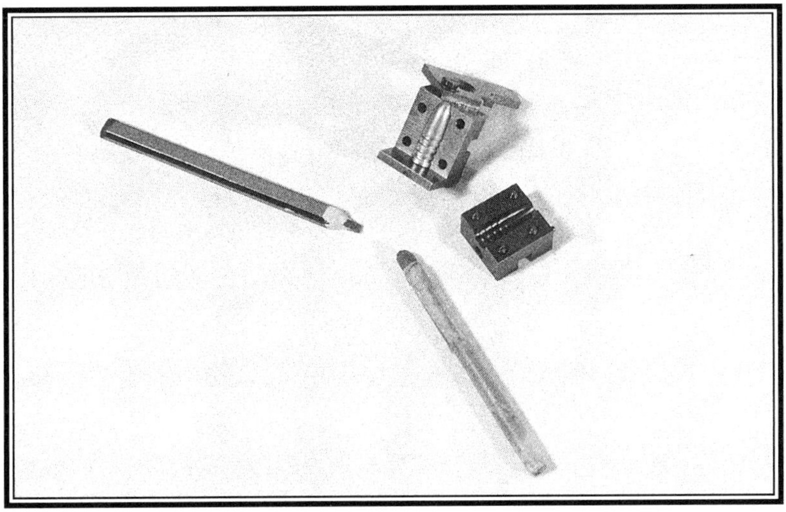

The set of nose-pour mould blocks was made by Charlie Canoll, shown with a beeswax lube pencil and a flat-lead carpenter's pencil. These latter two items should always be on hand when casting premium bullets.

10

How to Improve the Cutoff Plate

It does little good to maintain a constant temperature of the melt and bullet mould, and to cast a perfect bullet, only to destroy the gyroscopic balance of that bullet by leaving a ragged cavity in the base during the cutoff process. Although this seldom happens when using a soft alloy such as 30-to-1 lead and tin, it is a virtual promise when using antimonial alloys such as wheelweights with 2- or 3-percent tin.

Regardless of how little antimony is in the bullet alloy, its crystalline structure upon cooling seems to invite cavitation in the bullet base during the cutoff process. To make matters worse, this cavitation is never in the center of the base nor is it ever symmetrical. It always seems the sprue is partially cut by the sharp edge of the sprue hole or pour hole, and the remainder is torn away from the base of the bullet leaving a ragged off-center crater where it should be a smooth surface cut.

This is in part the result of cutting the sprue while the

metal is still soft, but cavitation is also aggravated by the blunt 45-degree angle on the cutting edge of the pour hole in the cutoff plate. The cutting edge should be at a much more acute angle to make a cleaner more complete cut when using the harder alloys.

When the sides of the sprue hole are cut at a 45-degree angle on a ⅛ inch thick cutoff plate as used on Lyman moulds, the top opening on the pour hole is a shy ⅜ inch diameter, barely large enough to accommodate the nozzle of the dipper. Although Lyman redesigned its dipper in 1995 and to some extent helped alleviate this problem, the nozzle on the dipper has a tendency to slip off the pour hole during the pouring process.

While there is no slippage problem with 3/16 inch thick cutoff plates as used on NEI, RCBS or SAECO moulds, the problem of the blunt 45-degree angle cutting edge and torn sprues still exists.

Why a 45-degree angle cutting edge was ever used is anyone's guess; but to the best of my knowledge, I have never seen a production mould, including those made by Marlin, Winchester and Ideal Manufacturing back in the 1800s, that was not cut at a 45-degree angle. And I suspect the old roundball moulds made for muzzle-loaders had the same angle. It was just something that got started and was never changed because it was never a problem when using the soft alloys that were so prevalent in those long bygone years.

If you want to improve the action of the cutoff plate on your bullet moulds, one of the ways to do it is to recut the angle on the pour hole, changing it from 45 to 35 degrees relative to the bottom of the cutoff plate.

You can do this or have someone else do it for you on a lathe. Grasp a short length of 2 inch diameter bar stock in a three-jawed chuck, face it off and drill a clearance hole in the center with a center drill. Lay your cutoff plate on the face of the bar stock with the pour hole centered over the clearance hole, and mark the center of the pivot screw hole. Also lay out two other holes to be drilled and tapped, one on each side of the cutoff plate. These are to keep the cutoff plate from moving during the machining process.

Drill and tap the three holes and secure the cutoff plate in place with machine screws. Place the jig back in the chuck and center the pour hole with a dial indicator. Swing the compound head around to 35 degrees, and with several light passes with the tool bit cut a new angle on the side of the pour hole. Do not enlarge the pour hole itself; merely cut a new 35-degree angle on the side of the pour hole right down to the edge itself.

If most of you are like me and do not own a lathe, or if you are not fortunate enough to have a friend who does, there is another way in which you can do the job yourself if you take your time and exercise a little care. Personally, I believe this does a better job than the lathe operation in that it puts a thinner cutting edge on the pour hole while maintaining sufficient metal strength behind the cutting edge to prevent distortion.

To do it the way I have done it myself, you must purchase a carbide burr having a ½-inch roundball cutter on a ¼-inch shank. These burrs are available from any industrial supply house and are relatively inexpensive. If you take care of the one you purchase, you can probably modify all the cutoff plates you will ever own without ever having to purchase another burr.

Unless you are modifying a brand new cutoff plate that has been blued, you should paint the pour hole with Dykem layout fluid. This will dry in a few seconds and will let you see just how far the cutting has progressed on the sides of the pour hole. This is important! You want to reshape the sides of the pour hole right down to the cutting edge without enlarging the pour hole itself.

Now get a 2-foot length of good, flat 2x4 wood. Lay the cutoff plate bottom side down with the protruding arm to your right in the center of the 2x4 and with a pencil mark the center of the pivot screw hole, the center of the pour hole and a third hole next to the arm on the side toward you. Using a 0.148 inch diameter drill, drill a hole on each of these locations 1¼ inches deep.

Place the cutoff plate on the 2x4 as you did before and run a 3⁄16-inch round head stove bolt into the holes you drilled for the pivot screw and behind the arm of the cutoff plate. This last one will provide a stop to prevent the cutoff plate from turning while the pour hole is being modified.

With the cutoff plate securely anchored, grasp the shank of the carbide burr in a brace (as in brace and bit) and liberally coat the cutting head of the burr as well as the pour hole with oil. Hold the burr firmly against the pour hole and start cranking the brace.

Do not apply too much pressure! The burr will cut very rapidly with only moderate pressure, so you want to stop frequently and check the progress of the cutting. This is where the Dykem layout fluid or blueing comes in handy, as it will show you at a glance just how close to the cutting edge the burr has done its work.

Don't be in a hurry. As you approach the end of the operation, use a light pressure on the brace to give a good finish to the final cut. Apply the oil liberally during the entire operation and keep cleaning the fine chips off the burr and the cutoff plate.

If you take your time in doing this operation with a carbide burr held in a brace, you will finish with a cutoff plate that makes a much smoother cut on the base of the bullet than the original plate.

There is so little metal to be cut away by the burr that if the burr is used in a power drill, I am afraid the cutoff plate would be ruined in a matter of seconds. I personally spend about half an hour on a single cutoff plate.

As mentioned earlier, using a carbide burr puts a thinner cutting edge on the cutoff plate than by machining it to 35 degrees. When I first started using a cutoff plate having the thin cutting edge, I was very careful in cutting off the first few sprues because I didn't know, and still don't, just how hard an alloy one can cut without deforming the cutting edge and possibly gouging the tops of the mould blocks. Though I believe there will be no problem at all in cutting Linotype sprues, as of this writing, the hardest alloy I have cut with the thin edge cutoff plate runs about 12 BHN. These have cut as easily as soft lead.

In talking to Ron Snover about the foregoing procedure, he advised it was his practice to modify the sprue hole with a ½-inch carbide burr or end mill to within about 1/64 inch of the cutting edge. At that point he finished the job with a 60-degree carbide burr held in his hand and manually rotated to give a fine 30-degree cutting edge without making it too thin.

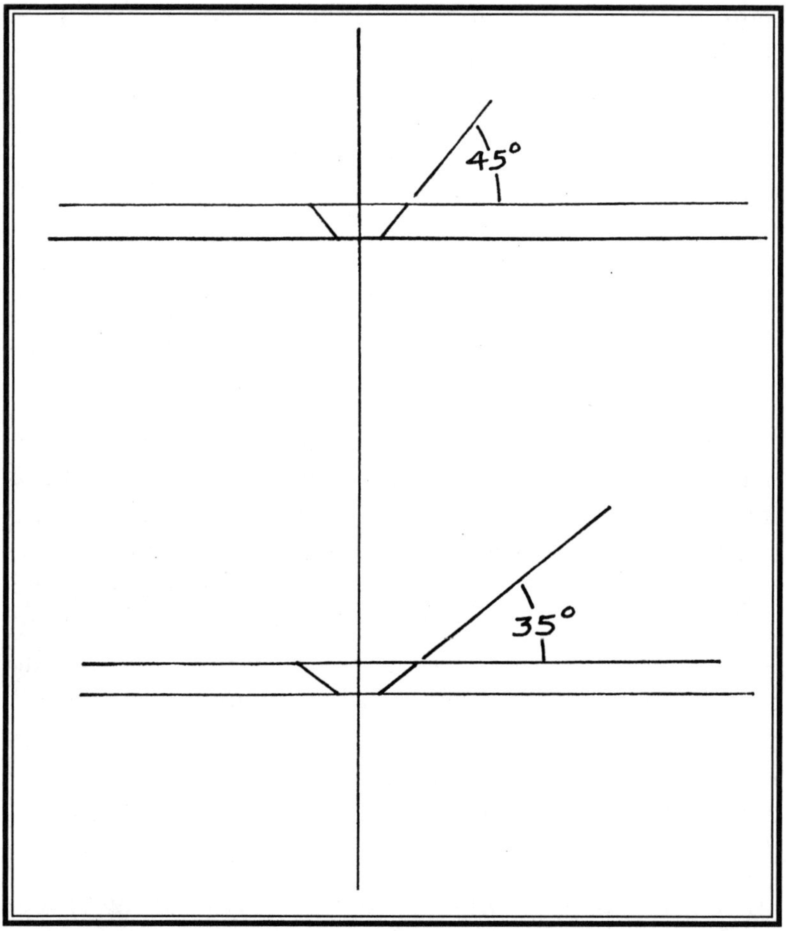

This sketch shows the difference in the cutoff angle on production cutoff plates (top) compared to that of a cutoff plate that has been modified on a lathe (bottom).

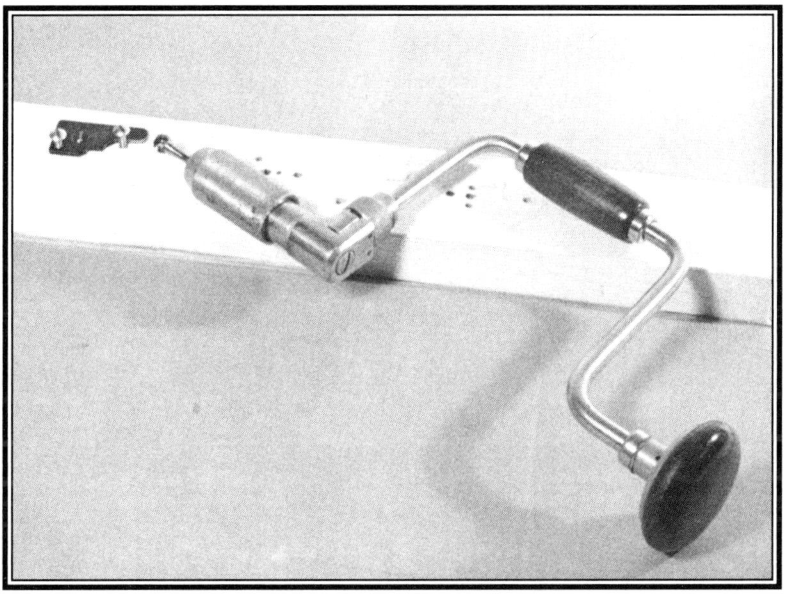

Equipment needed to modify the cutoff plate: hand brace, ½ inch diameter ball-shaped carbide burr and the cutoff plate securely held on a short length of 2x4 wood.

11

How to Make a Bullet Casting Board

Of all the components of the black powder rifle cartridge, the cast bullet is by far the most critical. When one has to be grouping his shots into minute and a half angle of accuracy or less out at 547 yards in order to be competitive, it is to his or her advantage to exercise the greatest care in casting and handling the bullets to be used.

For many years I have dropped my bullets onto a padding of old towels and, after a brief visual inspection, laid them side by side in the order cast in rows of 25 in a padded cardboard box. Later they were spot checked for weight variation and then stored in plastic pistol cartridge boxes with each bullet in its own individual compartment.

This system has worked well for me but is somewhat of a nuisance in that when the cardboard box has to be moved from one location to another with 16 or 17 pounds of bullets in it, the bottom sags, and I am afraid of dumping several hours of hard work onto the concrete floor. In addition, the folded towel in the bottom of the box never lies flat and is always home to bits of bullet metal debris from the casting operation. Of course, I clean the towel every time before starting a casting session, but the debris that collects during a casting operation clings to the fabric with the tenacity of beggar's lice. In short, it is not the ideal setup for cast bullets.

Very recently I designed what I refer to as a "bullet casting board" and had my brother Allan make one for me. This has proved to be one of the handiest casting accessories I have ever used, and I will never be without a bullet casting board again.

As shown in the drawing at the end of this chapter, this is basically a one-inch oak board having a .5-inch rim all around it and 25 shallow grooves located on .5-inch centers running lengthwise of the board.

I chose oak because I wanted the board to be heavy enough it could not accidentally be knocked off the chair or stool or whatever it might be resting upon during the casting process. The oak is heavy and gives plenty of rigidity, a far cry from the old cardboard box I had been using.

During the casting process, I still dump my bullets onto a soft pad placed across one end of the casting board. And after each bullet is cast and given a brief visual inspection, it is placed in one of the grooves beside the previous bullet cast. In this way bullets are kept in the order cast and can later be transferred to the pistol cartridge boxes in the same order. They can also be loaded and fired in the same order to help assure there will not be any abrupt change from one bullet to the next.

One of the fringe benefits of using such a casting board quickly became apparent when I started using a double cavity mould. Here I was easily able to segregate my bullets by mould cavity by placing each bullet from a given cavity alongside its predecessor in a row of its own. The same could also be done when casting with two moulds.

The bullet casting board also comes in handy if you segregate your bullets by weight after the casting process. In doing this operation, you can establish two, three or four rows of bullets by weight across the board until your entire run of bullets has been weighed and segregated, after which they can be boxed, labeled and stored for future use.

Although the drawing shows U-shaped grooves running the length of the board, the grooves can just as easily be square or V shaped. The important thing is that they be deep enough to keep the bullet from rolling and not so deep as to make it difficult to pick a bullet up with the fingers. Further, if the grooves are cut on .5-inch centers, the board will accommodate any caliber up to and including .45 and leave a slight space between each bullet. If you are going to be casting .50-caliber bullets, then the grooves should be on 9/16-inch centers.

Whatever bullet you cast, if you are a black powder cartridge rifle shooter, you surely realize the importance of perfection in the cast bullet. Although the bullet casting board will not help you cast perfect bullets, it will help you maintain that perfection after the bullets are cast.

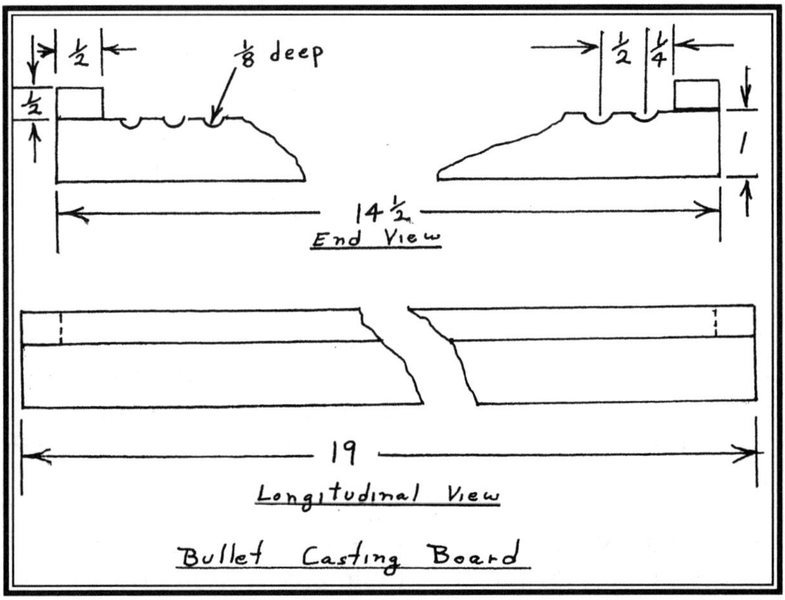

Dimension drawing of the bullet casting board made to handle bullets up to and including .45 caliber. Although round bottom grooves are shown, they can be square or V shaped.

The author's bullet casting board with bullets cast from a Pope-style nose-pour mould made by Dave Farmer of Colorado Shooter's Supply. Those are premium bullets on that board! Also, note the grooves in the board are ¼-inch square, not round as shown in the drawing.

12

The Custom Bullet Mould

I have written before that of all the custom mould makers who ever lived, the best of them are alive today. This is not to cast any stones at the likes of Harry Pope or some of the other old-timers who certainly made some wonderful moulds. But those who make bullet moulds today as a means of livelihood have better equipment for more precise machining and are called upon to make a far greater variety of designs at more specific dimensions than the old-timers were ever faced with. This is true because we have more shooters today, each with a different idea of the design and dimensions that will shoot best in his or her rifle. Where Harry Pope would make a mould for a bullet of his design to be fired in one of his barrels, the shooter of today submits a sketch or sample bullet with directions to cut a mould to cast a bullet exactly like this except . . .!

Make no mistake about it, the custom mould maker today can cut a mould that will cast within .0005 inch –

and usually closer – of your specifications. He can cut a mould that is usually .0005 or less out of round, and he can cut a mould so good you will be hard put to see the parting lines where the blocks come together along the side of the bullet. He can do this on a regular basis, time after time, because in this day of long-range precision black powder cartridge rifle shooting, those are the bullets top shooters demand.

So why not get a custom mould at the outset? Why waste time and money with production moulds when you know in the end, if you stick to the game, you will eventually get a custom mould cut to your specifications?

A custom mould in itself is no guarantee of precise accuracy at any range. In fact, there are many production moulds out there that give superb performance, and I believe the shooter just getting into black powder cartridge rifle shooting is far better off with a production mould from Lyman, RCBS, SAECO or NEI than an expensive custom mould that may or may not do any better.

If I were starting out today with a brand new .45-70 rifle and not having had any experience with black-powder loads, I would purchase one or two of the following production moulds:

- Lyman No. 457132 new style Postell
- Lyman No. 457125 roundnose military type
- RCBS No. 45-500-BPS (Part No. 82085)
- Redding-SAECO No. 745
- NEI No. 458-525-PB old-style Postell (No. 348A)

Although each of us may have a preference for bullet style or shape, the facts are at the velocities at which we are shooting, style or shape has minimal effect with trajectory or the so-called wind-bucking qualities. Any one of the foregoing bullets – if it is accurate in your rifle – will make a good silhouette bullet.

Cast your bullets of a 20-to-1 lead-tin alloy and try shooting them at first as-cast. That is, do not resize the driving bands unless you find an as-cast bullet will not fit inside a well-chamfered mouth of a fireformed unsized cartridge case. If this condition exists, put the bullet through the largest bullet resizing die available for that caliber. Try it for accuracy, and if you are not satisfied, go to the next smaller resizing die.

The object here is to determine what diameter driving bands shoot the best in your rifle. Despite what you may have read over the years, not all black powder cartridge rifles shoot their best with a groove diameter bullet or one that is .001 inch over the groove diameter of the barrel. In my own .45-70 silhouette rifles, I have been using bullets that measure .455 inch diameter over the driving bands for the past year and a half. This is .003 inch smaller than the groove diameter of the barrel!

Sometimes you will find that no matter what diameter the driving bands are, the bullet refuses to give good accuracy. Should that occur, check the nose diameter of the bullet and the bore diameter – across the tops of the lands – of your barrel. Again, despite all you may have read, not all barrels will digest a bullet having a bore-riding nose, i.e., a bullet whose nose diameter is the same as or a bit larger than the bore diameter of the barrel.

Although theory tells us a bullet with a bore-riding nose should give pinpoint accuracy because of its near-perfect alignment within the barrel, this is not always the case in a black powder cartridge rifle. In my own rifles in which the lands are narrower than the grooves, the diameter of the nose of the bullet just ahead of the front driving band must be at least .003 inch smaller than the bore diameter of the barrel. That is, for a bore diameter of .450 inch, I want a bullet having a nose diameter no larger than .447 inch. To make certain the nose diameter does not interfere with the bore diameter, I usually order my moulds to cast .445 inch just ahead of the front driving band.

One way to find all this out is to beg or borrow a few bullets of various dimensions from other shooters and try them. You may also have some friend who is handy with a lathe make you a nose sizing die with which you can size the nose of your bullets down a few thousandths to determine what size shoots the best in your barrel.

Only after you have done considerable experimentation and are certain of the best diameter over the driving bands and on the nose of the bullet should you go ahead with a custom mould for your favorite style bullet. Even then, it may not shoot up to your expectations. Or it may shoot better than you expected. I have had several custom moulds made by some of the best craftsmen who ever lived and, without exception, the bullets cast in these moulds were exactly as specified. Some shot exceptionally well, some did not. But whether or not the bullet shot well, every one of them was to my specifications, and the fact some did not shoot well had nothing to do with the quality of the mould.

I mention this with special emphasis because more than one custom mould maker has told me his biggest problem was cutting moulds for shooters who felt if the bullet did not shoot well, the fault was in the mould. In many cases, moulds that were cut to the customer's specification were returned as defective.

Know what you are doing when you order a custom mould, and be prepared to "eat" the cost of that mould if it doesn't perform exactly as you expected. Who knows, it may be just the mould you need for your next rifle!

After I determined the diameters needed to give good performance in my rifles, I applied these diameters to the old-style (pre-1974) Lyman Postell bullet number 457132 and had a custom mould cut by Tom Ballard and, later, another by Steve Brooks. Both give excellent performance.

As I have suggested in the foregoing paragraphs, I determined the optimum diameter of the bullet driving bands needed for best accuracy in my rifles by the trial and error method – try one diameter and if it doesn't work, try another. There is nothing wrong with this, and in many cases, it is the only way you will ever determine what diameters work best in your rifle. However, if you give some thought as to what goes on inside your rifle chamber at the instant of firing, then you can often closely predict ahead of time the optimum diameter of the driving bands for your bullet.

There are two vitally important things that happen at the instant of firing a black powder cartridge rifle loaded with a lead alloy bullet. First, the gases of the burning powder pressurize the inside of the cartridge case forcing the wall of the cartridge case outward against the wall of the chamber, thus sealing the chamber against gas leakage and releasing the grip or neck tension of the cartridge case on the bullet. If there is not room for the cartridge case to expand sufficiently to release the bullet, you will have erratic pressures and erratic muzzle velocity with a corresponding loss of accuracy.

Second, the suddenly released gases of the burning powder along with whatever powder has not yet burned slam into the base of your lead alloy bullet, thus bumping up or increasing the diameter of the bullet to partially fill the space just vacated by the expanding cartridge case. It is my opinion that if there is not sufficient clearance between the inside wall of the expanded cartridge case and the expanded or bumped-up bullet, you will again get erratic pressure, erratic velocity and a loss in accuracy.

With this in mind, take a cartridge case that has been fully fireformed in the chamber of your rifle and, using a dial caliper, measure the inside diameter of the mouth of the cartridge case. Make certain before measuring that the mouth of the cartridge case has been well chamfered so there is no burr or slight closing in of the mouth that would give you a false measurement.

The inside diameter of the mouth of a fireformed cartridge case will vary considerably depending upon the internal dimensions of the chamber and the wall thickness of the cartridge case. The .45-70 cartridge cases made by Remington and Winchester usually have a wall thickness of .010 inch whereas .45-70 cases made by Federal usually have a wall thickness of .011 inch.

This variation of internal chamber dimensions and wall thickness of cartridge cases can have a tremendous effect on the inside diameter of the mouth of a fireformed cartridge case. And the inside diameter of the mouth of a fireformed cartridge case can be vitally important relative to the optimum diameter of the driving bands on the bullet.

For example, when I use Remington cartridge cases in my .45-70 silhouette rifle, the inside diameter of a fireformed cartridge case measures about .459 to .460 inch, whereas the inside diameter of the Federal cartridge cases used in the same rifle measure about .457 to .458 inch, or about .002 inch smaller.

Over the years of using this rifle, I have found when I use Remington or Winchester cases, I should also use a lead alloy cast bullet having a diameter of no more than .4575 inch over the driving bands. In fact, I have had several custom moulds made to cast bullets of that diameter. Most of them give good accuracy. However, when I use Federal cartridge cases in that rifle, I use lead alloy cast bullets having a diameter of .455 inch over the driving bands. Steve Brooks made me a mould to cast a Postell bullet of that diameter and it, too, is superbly accurate.

The upshot of all this is if you carefully measure the inside diameter of the mouth of a fireformed cartridge case and then try a bullet that is .002 to .003 inch smaller, the odds are you will come pretty close to the optimum driving band diameter for your rifle with that particular make of cartridge case.

While I realize this may be contrary to all you have read recommending bullets that are .001 inch larger than the groove diameter of the barrel, or bullets that are the same diameter as the throat of the barrel, you have to remember in most cases those recommendations are for cast bullets being used in front of smokeless powder. They do not bump up or expand in diameter as do soft lead alloy cast bullets being used in the black powder cartridge rifle.

You should also remember hardness of alloy plays a vital part in all this. If you are using a bullet that is .001 to .003 inch smaller in diameter than the groove diameter of your barrel, that bullet has to bump up in order to seal the barrel and fully grasp the rifling. In order to bump up, it must be fairly soft – about 20-to-1 lead and tin. Conversely, if you use a bullet that is too soft, it can be distorted during the bumping-up process, thus destroying its potential for fine accuracy.

Finally, there are no cut-and-dried formulae that one can follow and be assured of minute-of-angle accuracy in the black powder cartridge rifle. The best we can do is try what has worked for others and hope it works for us. If it doesn't work as expected, do not be too hasty to condemn the method. Carefully analyze what you have done and try again with another bullet.

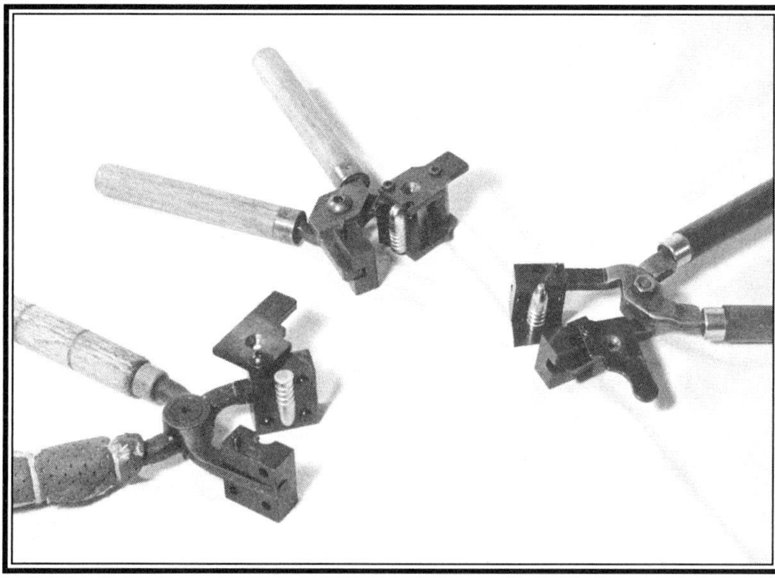

Three excellent custom bullet moulds: On the left is a base-pour mould made by Steve Brooks for casting the original Postell design bullet. Top center is a Pope-style nose-pour mould made by Dave Farmer for casting the Lyman 457125 roundnose bullet. On the right is a custom mould originally cut on Lyman blocks by Tom Ballard. Charlie Canoll later modified this mould to make it a nose-pour.

13

Nose-Pour Versus Base-Pour

Most black powder cartridge rifle matches are won with bullets from a base-pour bullet mould. If this were not true, virtually every custom mould maker out there would be advertising and making the nose-pour variety instead of the base-pour. They are not. Base-pour moulds are easier to make and easier to use than a nose-pour. It is that simple.

Over the years I have used both varieties of bullet moulds casting tens of thousands of bullets, most from the base-pour. Despite the fact in the past – before I learned what I know today – the bullets from the nose-pour moulds never seemed to give that extra bit of performance that made them worth the extra effort, I have to say without any qualification once you get a nose-pour mould casting properly, virtually every bullet you cast is of premium quality – and they look it!

There is no comparison between the base of a bullet from a nose-pour mould and one from a base-pour. The former is as smooth as a sheet of glass without a hint of imperfection, while the latter – though it may appear flat – shows evidence of the cutoff and sometimes evidence of more shrinkage around the cutoff than out by the rim of the rear driving band.

In order to see what I am talking about, take a dozen of your very best bullets cast in a base-pour mould and, using a 6-inch steel scale as a straightedge, place the edge of the scale across the center of the base while holding it up to a light. Does the scale touch the base all the way across? Or is there a high spot on the cutoff because the bullet shrunk endwise before the sprue was cut off? Is the cutoff centered and is it symmetrical? Or is the bullet slightly unbalanced?

If you have a good bullet cast from a nose-pour mould, go through the same straightedge exercise with it and notice the difference!

In the case of premium bullets cast in a base-pour mould, there is no question but what they will give superb accuracy regardless of the fact the straightedge shows they are not perfectly flat all the way across nor are they in perfect balance. Be that as it may, the very same bullet cast in a nose-pour mould with a base that is perfectly flat all the way across will give even better accuracy!

The difference in accuracy may not be much, and it may not be apparent until you get out to 300 yards and beyond. But it is there, just the same as the difference in the quality of the bases of the two bullets is there.

Because of the fact that when cooling, cast bullets shrink endwise as well as diametrically, there is no way to cast from a base-pour mould without having the cut-off protrude a few thousandths of an inch beyond the base of the bullet unless, of course, you create a cavity in the base of the bullet by not letting the sprue cool sufficiently. Cavitation is far more detrimental to accuracy than the slightly protruding cutoff.

I have used two different types of nose-pour bullet moulds, the Pope-style mould having a bottom plate that pivots with the cutoff plate and is connected to the cutoff plate by means of a .5 inch diameter spacer and a through bolt extending from the cutoff plate down through the spacer and screwing into the bottom plate. These are excellent moulds produced today by Colorado Shooter's Supply. Dave Farmer has made half a dozen of these moulds for me, and they all cast to the dimensions specified.

The other type of nose-pour mould has the bottom plate rigidly attached to one mould block. While at the present time no one advertises a mould of this type, I am certain there are those mould makers out there who have made such nose-pour moulds for themselves and others. My friend of more than 50 years, Charlie Canoll, has made a few of these moulds for me, and they work to perfection. In fact, it is not too difficult for a skilled craftsman to convert a base-pour mould to a nose-pour mould of this style.

Regardless of which style nose-pour mould one is using, you have to keep the melt and mould much hotter than when using a base-pour mould. This is because of the additional bottom plate and, in the case of the Pope-style mould, the spacer. Until that bottom plate and spacer get thoroughly heated to the proper temperature, the base of the bullet will be etched with fine wrinkles, and the edge around the base will not be sharp.

My usual casting temperature when using a nose-pour mould is around 840 to 850 degrees Fahrenheit. With the Pope-style nose-pour and this high a temperature, the base of the bullet sometimes has a tendency to adhere to the bottom plate. Thus when the bottom plate is swung aside to drop the bullet, there is sometimes a heavy smear of bullet metal that adheres to the bottom plate and mars the surface of the base.

Letting the bullet cool a bit longer to avoid this also cools the rest of the mould and promotes a rounded edge on the base, but treating the inner surface of the bottom plate with Lyman's Super Moly Spray Lube eliminates the problem. I expect that Mold Prep from NEI Handtools, Inc. would do as well. Whichever you choose to use, do not get any of it in the mould cavity or on the inner faces of the mould blocks. The best way to be certain of this is to put a bullet in the mould cavity and tightly close the handles while applying the Super Moly Spray Lube or Mold Prep to the bottom plate.

One other thing about the Pope-style nose-pour mould that can give you fits is the adjustment of the cutoff and bottom plates. In order to be absolutely certain the base of the bullet is square with the axis of the bullet, the cutoff plate and the bottom plate must be parallel. If they are not parallel, usually due to tightening the pivot screw too much, you will not get the accuracy from the bullet you should get because the base of the bullet will not be square with the axis.

Getting the two plates parallel and properly adjusted starts with a spacer of the proper length. To do this, remove the two plates from the mould blocks and remove the spacer from between the two plates. Thoroughly clean and dry the mould blocks, spacer and plates. With a good set of micrometers, measure the length of the spacer and the height of the mould blocks.

If the spacer is more than .0015 inch longer than the height of the mould blocks, place a sheet of 320 grit aluminum oxide paper on the flat surface, and working one end of the spacer in a figure-8 pattern, trim the length of the spacer so it is .0015 inch longer than the height of the mould blocks. You can go slightly longer than this if you want, but do not go any shorter as there must be some clearance between the mould blocks and the plates in order to prevent galling.

After you have the spacer ground to the correct length, reassemble the spacer between the two plates and slip the assembly over the mould blocks. Install the pivot screw down through the cutoff plate, mould block and into the bottom plate until it is just snug, no tighter.

Using a dial caliper, measure the distance between the two plates at the spacer and at the pivot end of the plates. If the plates are parallel, the measurement will be the same in both locations, and the plates should swing freely on the mould blocks. Install the locknut on the bottom end of the pivot screw and check everything again, making any final adjustment as necessary.

Once you get one of these Pope-style nose-pour moulds adjusted and casting properly, you will cast better bullets that will shoot more accurately than if they were cast in a base-pour mould. This I can promise you from my own personal experience.

The nose-pour mould with a rigid bottom plate attached to one of the mould blocks does not require the adjustment of the cutoff plate and bottom plate as does the Pope-style. Although I thought at first it would not drop bullets as well, such is not the case. Either style mould will cast equal quality bullets with just about the same amount of effort.

Very recently Charlie Canoll converted a custom base-pour mould of mine to nose-pour and, in so doing, has provided me with one of the most, if not the most, accurate bullets I have ever fired in a black powder cartridge rifle. For years I have used an original design Postell bullet in most of my match shooting, but with this new bullet from Charlie's converted mould, the accuracy is measurably better.

Originally this was a custom version of Lyman's Schmittzer bullet 457658 in that an additional driving band and grease groove had been added up front, making the bullet weigh about 508 grains. The mould cast good bullets, and the bullets gave good accuracy, though not as consistently good as my Postell bullet. Thus it was when Charlie told me he could convert the mould from base-pour to nose-pour, I was only too glad to let him try.

When Charlie makes such a conversion, the cutoff plate and the cutoff plate stop pin are removed from the blocks and a 3⁄16 inch thick steel bottom plate is rigidly secured to one block with two cap screws. The blocks are then turned end-for-end and what was the bottom end of the blocks is faced off in a lathe just far enough to make a 1⁄8 inch diameter opening into the nose end of the mould cavity. A new hole is drilled and tapped to accommodate the pivot screw, and a second hole is drilled to accommodate the cutoff plate stop pin.

Converting the five-band Schmittzer bullet mould in this fashion removed the needle sharp spitzer point from the bullet, replacing it with a 1⁄8 inch diameter meplat. It is my opinion – and this is only an opinion –

the combination of the small meplat and the much improved base on the bullets cast in this nose-pour conversion are responsible for the greatly improved accuracy of this bullet.

I strongly believe if you have a bullet from a base-pour mould that is or has given you superior performance in the past, that same bullet from a nose-pour mould will do even better. One of my future projects will be to take one of my favorite custom moulds for the original style Postell bullet and have it converted to nose-pour. That, I believe, might be the ultimate!

Another view of the Pope-style nose-pour mould made by Dave Farmer of Colorado Shooter's Supply for casting the Lyman 457125 bullet.

14

The Future

Although I have been casting bullets for the past 50 years, and during those years did a lot of experimentation, it has only been since I got into competitive shooting with the black powder cartridge rifle that I focused my efforts toward precise accuracy with the cast bullet. It was during the course of these efforts I learned that a good share of what has been written relative to cast bullets over the past 80 or 90 or 100 years has nothing to do with the black powder cartridge rifle. In short, we who are striving for precise accuracy with cast lead alloy bullets in the black powder cartridge rifle are standing in an uncultivated field.

In these few short years of revival of black powder cartridge rifle competition, we have barely scraped the surface of the accuracy mother lode. This, I believe, is because we have relied too much on our previous experience with cast bullets used in high-powered rifles and handguns, both entirely different ball games.

Yet, we are gaining. The quality of rifles and barrels and sights has reached the peak of perfection. Winning scores are much higher than just eight or 10 years ago indicating shooters, too, have improved in their knowledge of wind, mirage and shooting technique. Even our black powder has improved to the point where I believe it is comparable to some of the best produced in the late 1800s.

So if we are to push our average scores higher than those of today, we must look to improvement in the cast bullet. We must cast more perfect bullets than we cast today and those, I believe, will come from a nose-pour mould. I believe the nose-pour mould is the mould of the future for black powder cartridge rifle shooters. And I believe the top black powder cartridge rifle shooters and the riflesmiths who build their rifles will more closely study the relationship between chamber dimensions, cartridge case wall thickness and the optimum diameter of the driving bands of the bullet being used.

This, I believe, will come to pass, and I will do my utmost to be a part of it.

Good shooting to all of you!

Books by Paul Matthews:

Forty Years with the .45-70

The Paper Jacket

Loading the Black Powder Rifle Cartridge

Shooting the Black Powder Cartridge Rifle

How-To's for the Black Powder Cartridge Rifle Shooter

Cast Bullets for the Black Powder Cartridge Rifle

Wind Drift and Deceleration of the Cast Bullet
at Black-Powder Velocities

Casting Premium Bullets for the
Black Powder Cartridge Rifle

Black Powder, Pig Lead and Steel Silhouettes

More How-To's for the Black Powder
Cartridge Rifle Shooter

Dear Folks

What You Should Know About Bullet Lubricants
for the Black Powder Cartridge Rifle

Things You Should Know About Shooting the
Black Powder Cartridge Rifle